Hanan S. Badawy

Rudist, Coralline Sponge, Gastropod in El-Hassana Protectorate, Egypt

Hanan S. Badawy

Rudist, Coralline Sponge, Gastropod in El-Hassana Protectorate, Egypt

Growth of Cretaceous Tethys Realm Population, Facies Hierarchy, Depositional Model

LAP LAMBERT Academic Publishing

Impressum / Imprint
Bibliografische Information der Deutschen Nationalbibliothek: Die Deutsche Nationalbibliothek verzeichnet diese Publikation in der Deutschen Nationalbibliografie; detaillierte bibliografische Daten sind im Internet über http://dnb.d-nb.de abrufbar.
Alle in diesem Buch genannten Marken und Produktnamen unterliegen warenzeichen-, marken- oder patentrechtlichem Schutz bzw. sind Warenzeichen oder eingetragene Warenzeichen der jeweiligen Inhaber. Die Wiedergabe von Marken, Produktnamen, Gebrauchsnamen, Handelsnamen, Warenbezeichnungen u.s.w. in diesem Werk berechtigt auch ohne besondere Kennzeichnung nicht zu der Annahme, dass solche Namen im Sinne der Warenzeichen- und Markenschutzgesetzgebung als frei zu betrachten wären und daher von jedermann benutzt werden dürften.

Bibliographic information published by the Deutsche Nationalbibliothek: The Deutsche Nationalbibliothek lists this publication in the Deutsche Nationalbibliografie; detailed bibliographic data are available in the Internet at http://dnb.d-nb.de.
Any brand names and product names mentioned in this book are subject to trademark, brand or patent protection and are trademarks or registered trademarks of their respective holders. The use of brand names, product names, common names, trade names, product descriptions etc. even without a particular marking in this work is in no way to be construed to mean that such names may be regarded as unrestricted in respect of trademark and brand protection legislation and could thus be used by anyone.

Coverbild / Cover image: www.ingimage.com

Verlag / Publisher:
LAP LAMBERT Academic Publishing
ist ein Imprint der / is a trademark of
OmniScriptum GmbH & Co. KG
Bahnhofstraße 28, 66111 Saarbrücken, Deutschland / Germany
Email: info@lap-publishing.com

Herstellung: siehe letzte Seite /
Printed at: see last page
ISBN: 978-3-659-81431-0

Copyright © 2015 OmniScriptum GmbH & Co. KG
Alle Rechte vorbehalten. / All rights reserved. Saarbrücken 2015

Table of Contents	
Preamble	2
Part One: Stratigraphy	2
1.1 Synonym	4
1.2 Reference section and thickness	4
1.3 Stratigraphic boundaries	5
1.4 Lithologic aspect and succession	5
1.5 Fossil content and geologic age	11
1.6 Distribution and correlation	12
Part Two: Petrography	13
2.1 Oyster shell grainstone	13
2.2 Bioclastic rudist floatstone	14
2.3 Gastropod shell packstone-grainstone	14
2.4 Bioclastic peloidal packstone	16
2.5 Laminated peloidal foraminiferal grainstone	17
2.6 Bioclastic planktonic foraminiferal wackestone-mudstone	18
2.7 Terrigenous bioclastic wackestone	19
2.8 Bioclastic whole echinoderm fossil wackestone	19
2.9 Bioclastic ostracodal wackestone-mudstone	20
Part Three: Facies Hierarchy and Depositional Model	20
3.1 Facies A: Bioturbated/massive skeletal wackestone-mudstone	20
3.2 Facies B: Well-bedded/bioturbated skeletal-peloidal packstone	23
3.3 Facies C: Massive/laminated oyster shell grainstone (Oyster Hash)	25
3.4 Facies D: Massive/bioturbated gastropod skeletal packstone-grainstone (Gastropod-shell beds)	27
3.5 Facies E: Massive rudist boundstone-bafflestone	28
3.6 Facies F: Laminated/bioturbated planktonicforaminiferal wackestone-mudstone	34
3.7 Facies development and depositional model at El-Hassana dome	36
Part Four: Concluding Remarks	37
References	39
Acknowledgments	

Preamble

El-Hassana dome looks like an open museum that keeps a full record of ancient life, climate and environment during the Late Cretaceous time span that lasted for about 35millions years ago. Owing to the all features of El-Hassana dome, it is one of the famous and attractive geological sites in Egypt, and it deserves to be one of the protectorate areas in Egypt. It has been declared by Prime minister as protectorate since 1989.

The landscape and geological framework of El-Hassana dome protectorate imitate its distinguished geologic history. The area represents one of the Syrian arc elements with wrench tectonic deformation style. It is the sole location near Cairo where the exposed rock units form a part of the Upper Cretaceous surface section (Fig. 1a&b) relative to the unexposed counterpart in vast territory of north Western Desert. It locates at Abu Roash area on Cairo-Alexandria desert road, about 8Km far from Giza Pyramids (Fig. 1a&b). El-Hassana forms conspicuous round dome-shaped hill of $1km^2$ area and with the highest point reaches up to 149m above sea level. Several successive concentric ridges and gullies surround the dome. Economically, the equivalent rock units of the Upper Cretaceous in the north Western Desert oil fields are potential source rock and good reservoirs for oil and natural gas. Fascinating population of the Tethyan Cretaceous rudist, coralline sponge and gastropods are best preserved in the area.

1. Stratigraphy

The considered rock unit is the fourth member of Abu Roash Formation consisting of limestone, marl and calcareous shale which yield, in some places, abundant rudist shells. It exhibits a contrasting dark tone relative to the encompassing white cliff forming' limestone-dominated members and is characterized by its enrichment with *Trochactaeon salomonis* (Fraas), *Nerinea requieniana* d'Orbigny and *Millestroma nicholsoni* Gregory.

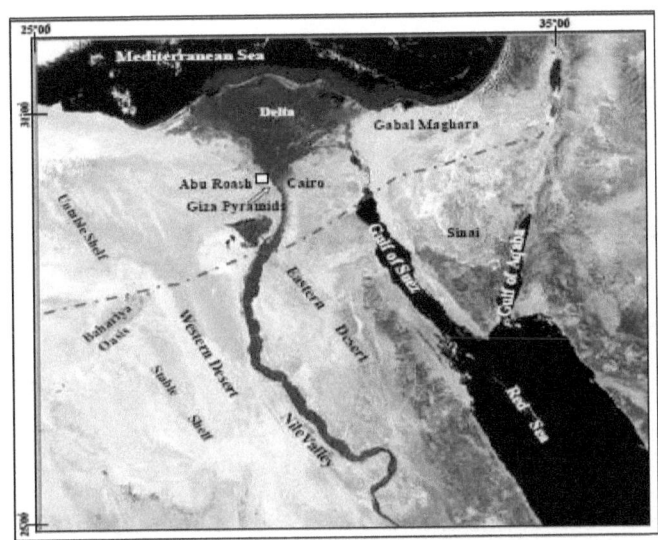

Fig. 1a: Satellite image of Egypt showing the location of Abu Roash area in North Western Desert.

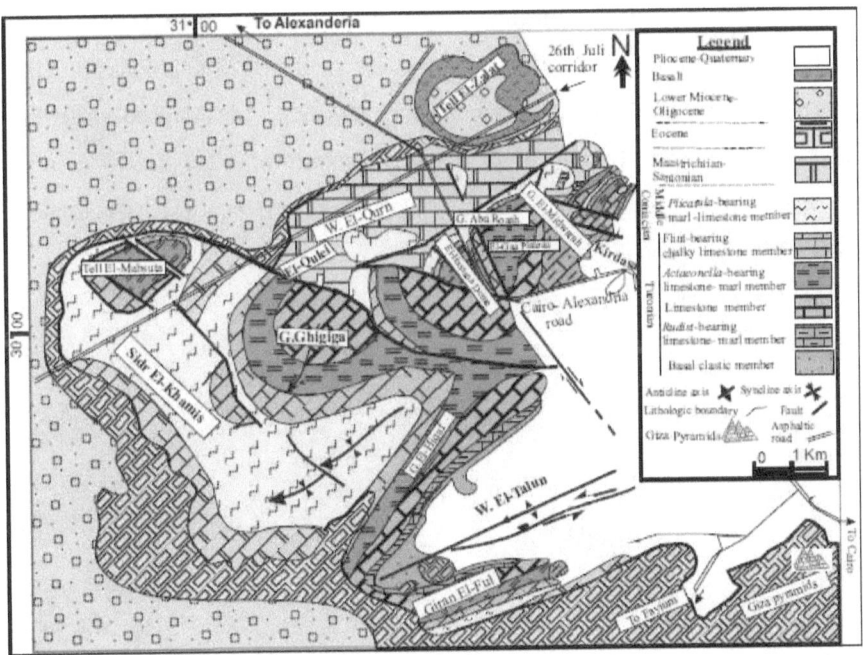

Fig.1b: Geologic map of Abu Roash area (modified after Faris, 1948; Sehim, 1986 and Badawy, 2003) showing the distribution of rock units in Abu Roash area.

1.1 Synonym

The *Actaeonella*-bearing limestone unit was treated under the following names:

a. "*Actaeonella-Nerinea* Limestone" (δ of Beadnell, 1902 and Osman, 1954a & b).

b. "*Actaeonella* Series" (T_3 of Faris, 1948; Omara, 1953; Jux, 1954; Said, 1962 and Hataba and Ammar, 1990).

c. "Abu Roash C" (Sehim, 1986 and Abdel Khalek *et al*., 1989).

d. *Actaeonella*-bearing limestone-marl member (Badawy, 2003 and Abu Khadrah *et al*., 2014)

1.2 Reference section and thickness

This unit is best exposed and measured at El-Hassana dome, where it flanks a core formed of the limestone member (Figs. 1b &2). Another reference section is described from the outcrops forming the western flanks of El-Gaa Plateau (Fig. 1b). The *Actaeonella*-bearing limestone-marl member had assigned different thickness that swings between 27m to 70m (e.g. Beadnell, 1902; Faris, 1948; Jux, 1954 and Abdel Khalek et al., 1989). At the reference sections, its compiled thickness measures about 37m.

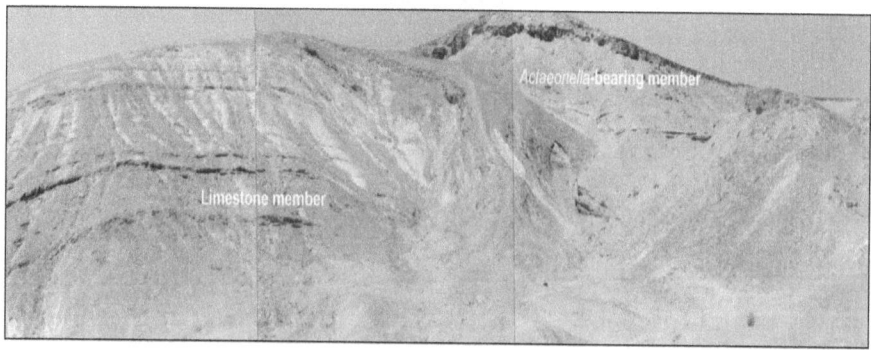

Fig.2: Panoramic view of El-Hassana dome. The core is formed of the white limestone member and the western flanks formed of *Actaeonella*-bearing limestone-marl member that displays a distinct yellowish brown tone and slope-forming nature.

1.3 Stratigraphic boundaries

The *Actaeonella*-bearing limestone-marl member is encompassed between two limestone dominated members the "flint-bearing chalky limestone" above and the "limestone member" below. Distinct yellowish green fissile shale demarcates its contact with the encompassing members.

1.4 Lithologic aspect and succession

The lithologic succession of this member has a rather dark tone (gray, yellowish brown and yellowish green) and slope-forming nature (Fig. 2) contrasting with the encompassing white and cliff-forming limestone members. The member can be easily subdivided into three distinct units being from base to top (Fig.3a&b):

- **Lower limestone-shale unit**
- **Middle limestone unit**
- **Upper limestone-shale unit**

The lower limestone-shale unit is best developed and exposed at El-Hassana dome where it attains about 19m thick. Further northeastwards, at the outcrops near Abu Roash Village, this unit remarkably thins in thickness to less than 7m. It consists mainly of glauconitic argillaceous limestone grading to calcareous shale and intercalating with thin- to thick-beds (5-25cm thick) of bioturbated fossiliferous limestone, flaggy bedded sandy limestone and oyster hashes (Fig. 4a). The glauconitic argillaceous limestone and shale are yellowish brown to yellowish green, slope-forming, highly gypsiferous, saliferous and yield abundant bivalves and gastropods (Fig. 4b).

The middle limestone unit is the gastropod-bearing stratigraphic interval characterizing this member. Everywhere in the study area, this unit is formed of thick- to very thick- either massive to faintly laminated or bioturbated grayish white limestone beds containing disoriented whole shells of *Trochactaeon salomonis (Fraas)* and *Nerinea requieniana d'Orbigny* (Fig. 4b & 5a).

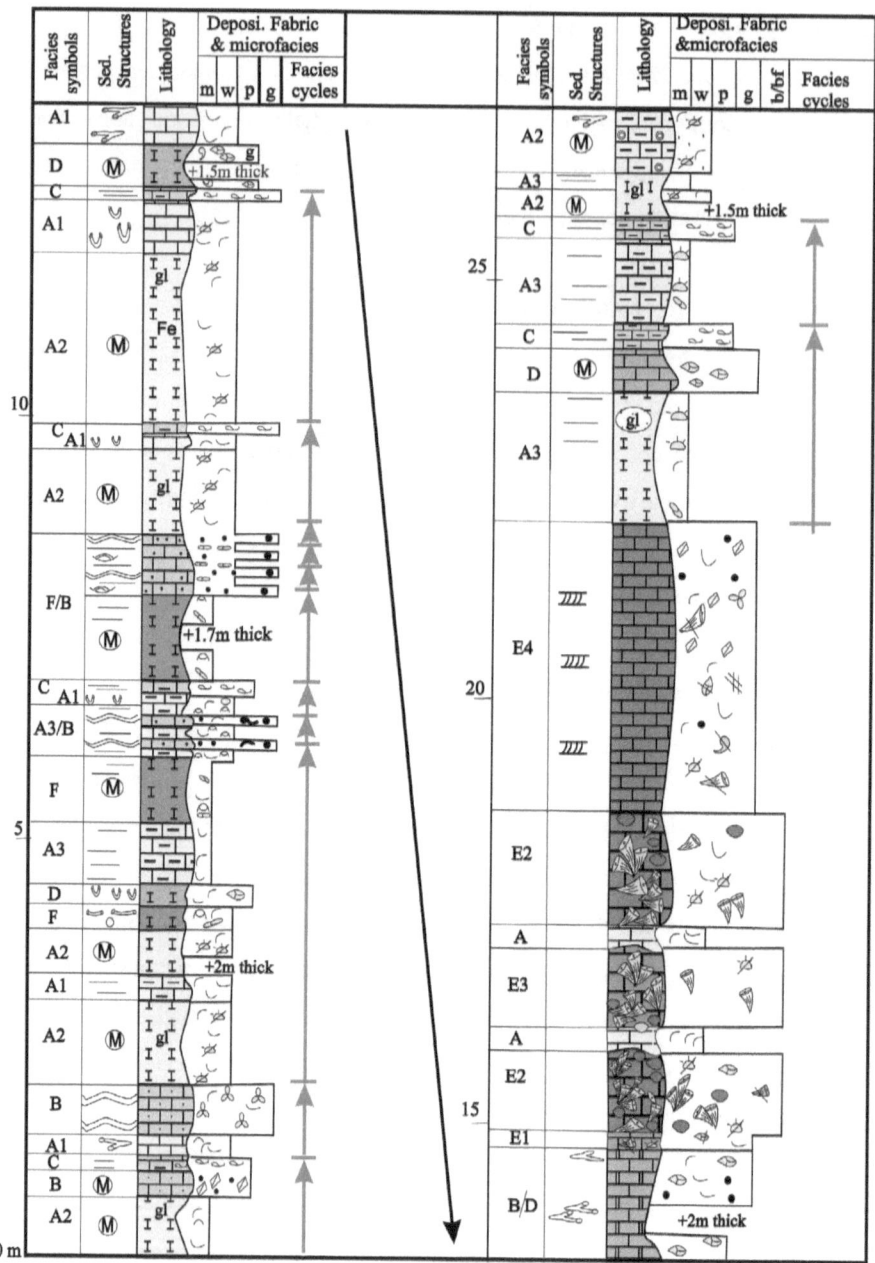

Fig.3a: Stratigraphy and sedimentary facies of *Actaeonella*-bearing limestone-marl member in Abu Roash area (modified after Badawy, 2003).

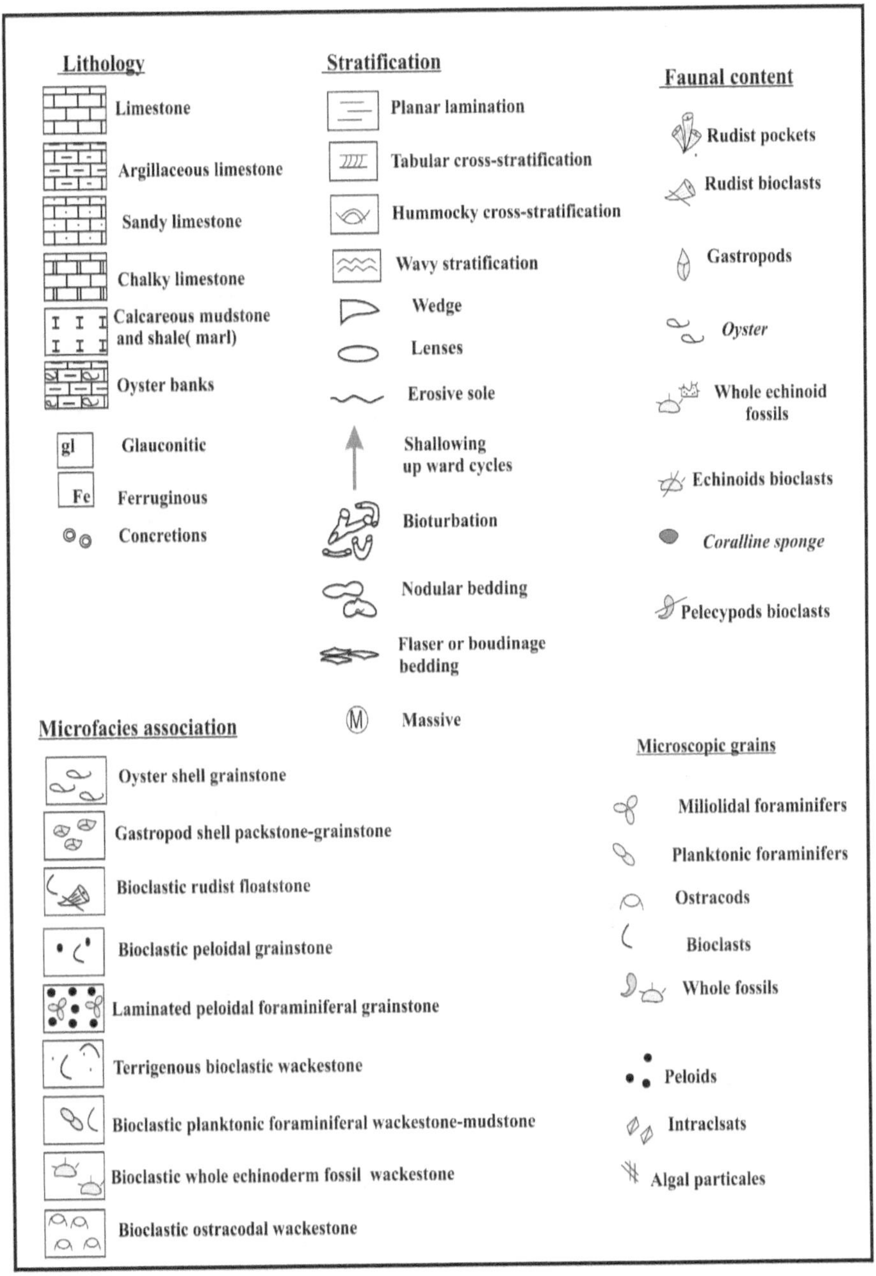

Lithology

- Limestone
- Argillaceous limestone
- Sandy limestone
- Chalky limestone
- Calcareous mudstone and shale (marl)
- Oyster banks
- gl — Glauconitic
- Fe — Ferruginous
- Concretions

Stratification

- Planar lamination
- Tabular cross-stratification
- Hummocky cross-stratification
- Wavy stratification
- Wedge
- Lenses
- Erosive sole
- Shallowing up ward cycles
- Bioturbation
- Nodular bedding
- Flaser or boudinage bedding
- Massive

Faunal content

- Rudist pockets
- Rudist bioclasts
- Gastropods
- Oyster
- Whole echinoid fossils
- Echinoids bioclasts
- Coralline sponge
- Pelecypods bioclasts

Microfacies association

- Oyster shell grainstone
- Gastropod shell packstone-grainstone
- Bioclastic rudist floatstone
- Bioclastic peloidal grainstone
- Laminated peloidal foraminiferal grainstone
- Terrigenous bioclastic wackestone
- Bioclastic planktonic foraminiferal wackestone-mudstone
- Bioclastic whole echinoderm fossil wackestone
- Bioclastic ostracodal wackestone

Microscopic grains

- Miliolidal foraminifers
- Planktonic foraminifers
- Ostracods
- Bioclasts
- Whole fossils
- Peloids
- Intraclsats
- Algal particales

Fig.3b: Symbols of lithology, stratification, fauna, microscopic grains and microfacies association used in Figure 3a (modified after Badawy, 2003).

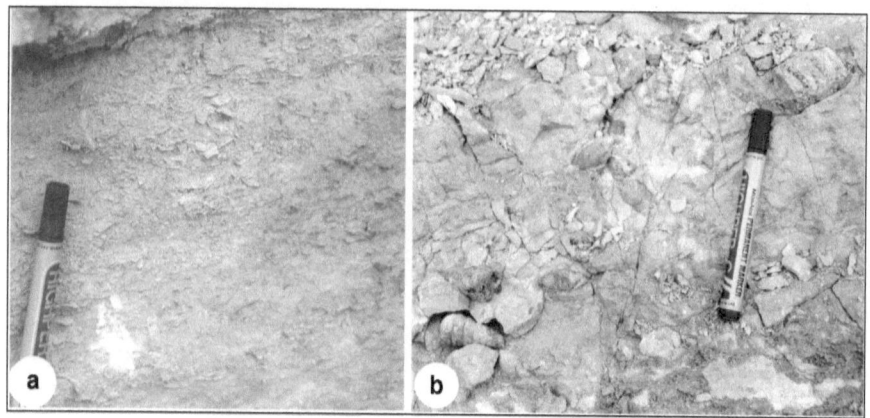

Fig.4: (a) Close up view of yellowish brown to yellowish green slope-forming glauconitic argillaceous limestone yielding abundant bivalves (mainly oysters), lower limestone-shale unit, *Actaeonella*-bearing limestone-marl member. (b) Close up view of well-preserved shells of *Nerinea requieniana* and *Trochactaeon salomonis* in the basal fossiliferous beds of gastropod-bearing interval, *Actaeonella*-bearing limestone-marl member.

At the reference section El-Hassana dome; this limestone unit starts with 2-3m thick of highly bioturbated chalky limestone beds that flooded with *Trochactaeon salomonis* (Fraas). These beds are followed up by another interval of limestone (about 8m thick) being overcrowded with large-sized rudists of *Durania arnaudi* (Choffat) that reach in length to about 30cm and in diameter up to 15cm (Fig. 5b). They are internally recrystallized but the external wall still preserves its distinct radial ornamentation. These robust shells are almost compressed along their long axes. They are colonized in three to four irregular beds (1-2m thick), which are separated from each other by discontinuous layers (10-20cm, thick) of white bioclastic limestone containing skeletal fragments of rudists and echinoderms. In the basal and uppermost rudist beds, the *Durania* shells associate with the coralline sponge *Millestroma nicholsoni* Gregory (Fig. 5c), possessing hemispherical heads with surfacial nodes and mamillons (Pl. 5d). The sponge heads are more abundant and larger (up to 30cm in size) in the upper most *Durania* bed. The millestromid and *Durania* buildup is overlain and flanked (in NW direction) by faintly cross-bedded and firmly cemented fossiliferous limestone beds (about 3m, thick) (Fig. 6). These beds are consisting of diversified bioclasts and peloids.

Fig.5: (a) Middle gastropod-bearing limestone unit of the *Actaeonella*-bearing member characterized by well- thick to very thick massive bedding with disoriented gastropod shells and scattered hemispherical heads of coralline sponge. (b) Limestone beds overcrowded with large sized rudist shells of *Durania arnaudi* reaching in length to about 30cm and in diameter up to 15cm, middle limestone unit, *Actaeonella*-bearing limestone-marl member, El-Hassana dome. (c) Heads of the coralline sponge *Millestroma nicholsoni* (arrows) scattered in the basal *Durania* shell beds, middle limestone unit of the *Actaeonella*-bearing limestone-marl member, El-Hassana dome. (d) Close up view of isolated large silicified heads of the coralline sponges *Millestroma nicholsoni* (arrows) preserved in their growth position. Notice the hemispherical form of these heads and their surfacial nodes and mamillons, middle limestone unit of the *Actaeonella*-bearing limestone-marl member.

The rudist and millestromid limestone interval has a very local distribution in the study Abu Roash area. Its best occurrence seems to be restricted only to El-Hassana dome, even at this locality, such interval extends only for about 133.6m (Hamza, 1993). Northwestwards it changes to cross-bedded fossiliferous limestone beds, while in southeast direction, it is formed of about 2m thick limestone being crowded only with the millestromid coralline sponge *Millestroma nicholsoni* Gregory (Fig.6).

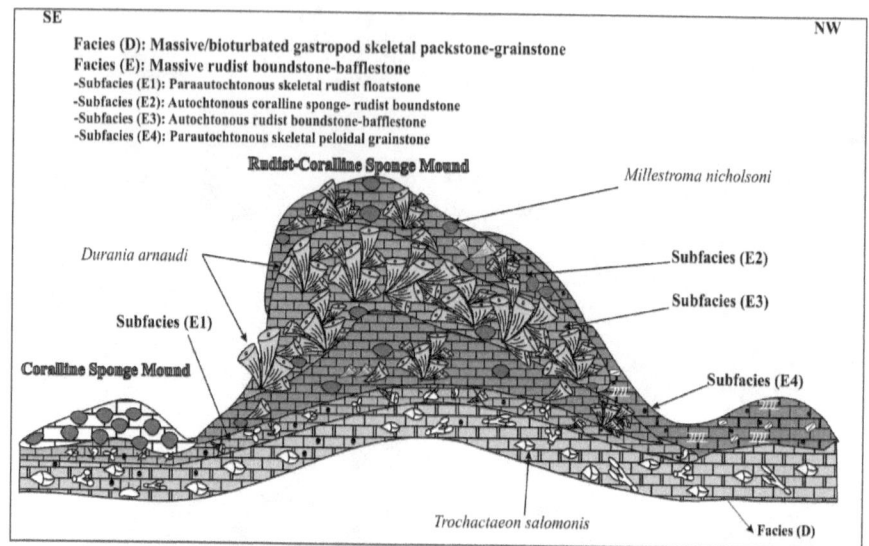

Fig.6: Schematic cross-section showing local growth and accumulation f rudist and coralline sponge buildup (massive rudist boundstone-bafflestone facies E and its subfacies E1-E4) forming mound and the southeast change to local buildup of the coralline sponge mound, middle limestone unit, El Hassana dome (Modified after Hamza, 1993; Abdel-Gawad, 2001; Badawy, 2003).

Northeastward of El-Hassana dome, to the western scarp of El-Gaa Plateau, the middle limestone unit of the study member attains an average thickness of about 10m.

The unit consists entirely of thick- to very thick- (0.5-1m) highly bioturbated limestone beds overcrowded with disoriented shells of *Trochactaeon salomonis* (Fraas) and *Nerinea requieniana* d'Orbigny. The uppermost beds commonly terminate with isolated large silicified heads of the coralline sponge, *Millestroma nicholsoni* Gregory being preserved approximately in its life position (Fig. 5a & c & d). Very local and isolated patches of the large sized *Durania arnaudi* (Choffat) are also recorded for the first time in some of the upper beds by Badawy (2003). Further northeastwards; near Abu Roash Village, the outcrops of the middle limestone unit reach in thickness, up to 18m, consisting almost entirely of very thick- massive to slightly bioturbated grayish white limestone beds. In these beds, a proportionally marked drop in the amount of the gastropod shells occurred. Only scattered shells are observed in the basal and upper beds of the unit. Also, there is no any trace of the rudist *Durania* shells, while the coralline sponge *Millestroma nicholsoni* Gregory exists in local isolated patches terminating this unit.

The upper limestone-shale unit attains about 7m thickness and is best outcropped at the western flank of El-Hassana dome, where it underlies the tilted chalky limestone beds of the overlying flint-bearing chalky limestone member. It closely resembles the lower limestone-shale unit, and consists of yellowish gray to green calcareous shale inter-bedded with ledge-forming highly fossiliferous glauconitic argillaceous limestone. The calcareous shale is slope-forming, fissile on the weathered surface, highly dissected with gypsum and salt veinlets, and yields abundant delicate bivalve shells. Near the base, the calcareous shale encloses discontinuous lenses of yellowish green glauconitic sandstone. The glauconitic argillaceous limestone interbeds (80cm up to 1m, thick) are hard, occasionally concretionary and very fossiliferous with oysters, some gastropods, echinoids and scattered shark teeth.

1.5 Fossil content and geologic age

There is a common acceptance in most previous studies that this unit belongs to Turonian age relying solely on the above-mentioned biostrome of *Durania arnaudi* (Choffat), *Trochactaeon salomonis (Fraas)* (Fig. 7 a-c) and *Nerinea requieniana d'Orbigny* (Fig. 7 d & e) (e.g. Beadnell, 1902; Faris, 1948 and Jux, 1954). In Sinai the *Durania-Trochactaeon* biostrome is overlain by flinty limestone that contains the typical Late Turonian ammonite *Coilopoceras requienianum* (Abdel-Gawad 1999 & 2000). However, *Durania arnaudi* (Choffat) is considered to be of probable Coniacian age by De Castro and Sirna (1996).

They followed the opinion of Hataba and Ammar (1990) in dating the "*Actaeonella* and *Durania*"-bearing beds to Coniacian age based on their collection of ostracods and forams including: *Hutsonia ascalapha* Bold, *Ovocytheridea symmetrica* Reyment, *O. producta* Grekoff, *O. apiformis* Reyment, *Haplophragmoides eggeri* Cushman, *H. gracilis* Said and Kenawy, *Ammobaculites* sp., *Discorbis turonicus* Said and Kenawy and *Ceratobulimina aegyptiaca* Said and Kenawy. However, this assemblage is not index for Coniacian and is repeatedly reported in Turonian deposits by many authors (e. g. Tewfik and Ebeid, 1975; El Shinnawi and Sultan, 1975; Wasfi et al., 1986 and Andrawis, 1990).

The present authoress agrees with Late Turonian assignment where this member and the overlying "flint-bearing chalky limestone" member rest below marl and limestone beds yielding the typical Coniacian ammonites *Tissotia tissoti* (Bayle), *Metatissotia fourneli* (Bayle) and *Heterotissotia neoceratites* Péron.

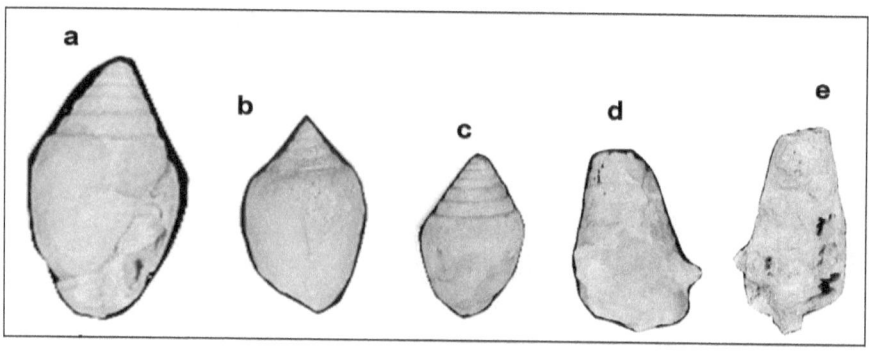

Fig. 7: (a-c) *Trochactaeon salomonis* (Frass). (b) apertural view, (c&d) abapertural views. Turonian, *Actaeonella*-bearing limestone-marl member, X0.5. (d&e) *Nerinea requieniana* d' Orbigny. Turonian *Actaeonella*-bearing limestone- marl member, X0.5.

1.6 Distribution and correlation

In addition to the reference section, the member forms the summit of Gabal El-Ghigiga and is recorded at Giran El-Ful, Gabal El-Hiqaf and Tel El-Mabsuta. It is correlated with upper part of Wata Formation in Sinai (Abdel-Gawad, 1999) and to the Abu Roash C member (Abdel Khalek et. al., 1989 and Hataba and Ammar, 1990).

2. Petrography

The detailed petrographic and microfacies analysis of the mound and associated sediments revealed the following microfacies types:

2. 1 Oyster shell grainstone

The microfacies forms several oyster hashes that vary in thickness from 5 to 50 cm. It consists mainly of whole and fragmented oyster shells (> 40% of the total rock composition) with subordinate amounts of ostracod carapaces (Fig.8). Most of the oyster shells display a preferred orientation (Fig.8) and still preserve their original foliated wall microstructure. However, some shells are either recrystallized (calcitized) or occasionally micritized by endolithic microbes (Fig.8).

The oyster hash rock is grain-supported and moderately sorted grainstone. It displays commonly a bimodal size of framework components including fine sand (peloids and terrigenous grains) and rudite grades (oyster shells) that are welded together with fine granular sparry calcite.

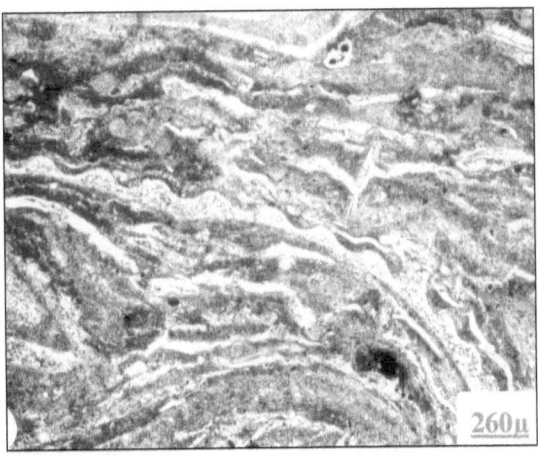

Fig.8: Oyster shell grainstone consists of whole and fragmented recrystallized oyster shells and subordinate ostracods. Notice the preferred orientation of the shells and the occasional micritization by endolithic microbes, Polarized light (P.L).

2. 2 Bioclastic rudist floatstone

The rock of this microfacies forms one irregular lenticular bed (1-2m, thick) The rock is a coarse-grained, open-packed, poorly sorted with floatstone packstone- grainstone matrix. It consists essentially of rudist shells (50% of the rock composition) mixed with echinoid plates and spines. Its grain-supported matrix consists of benthonic foraminifers of miliolids and cuneolinids, fragments of inoceramids, ostracod valves, crinoid fragments and peloids set in argillaceous or ferruginous micrite. The essential rudist particles are mostly unworn and still preserve their diagnostic wall microstructure of rectangular grid voids (Fig. 9). The rectangular voids are almost occluded with micrite and/or orthospar being occasionally dolomitized or partially replaced by spherulatic authigenic quartz. The other bioclasts of this rock type are either partially micritized (e.g. miliolids and algal particles) or enlarged by a syntaxial calcite overgrowth (e.g. echinoids).

Fig. 9: Well-preserved rectangular grid voids characterizing the wall microstructure of the rudist shell. It is occluded with micrite and orthospar, crossed Nicole (C.N).

2. 3 Gastropod shell packstone-grainstone

The rock comprising this microfacies forms the middle limestone unit and some layers of the lower and upper limestone-shale units (Fig. 3a). The beds of this microfacies range in thickness from 0.3cm up to 1m and are almost internally massive to slightly bioturbated. *Actaeonella salomonis* and *Nerinea requieniana* are the main gastropod shells building this rock type.

In thin section, the rock is grain-supported, moderately sorted and of rudite size. It consists almost entirely of whole fossils and some fragments of gastropods (~35% of the total composition) with scattered bivalves (Fig. 10a-c). Almost all the original gastropod shells are completely dissolved and their mouldic and whorl cavities are occluded commonly with pelletal and glauconitic micrite enclosing silty-sized bioclasts (Fig. 10b-e). The mouldic cavities of some shell wall, shell columella as well as some inter- shell voids are filled with fine to coarse-crystalline granular orthospars (Fig. 10b-e). The other inter-particles pore spaces are occupied with a micritic matrix being in parts glauconitic and in others with organic matter. Scattered fines to coarse crystalline dolomite rhombs are also recorded.

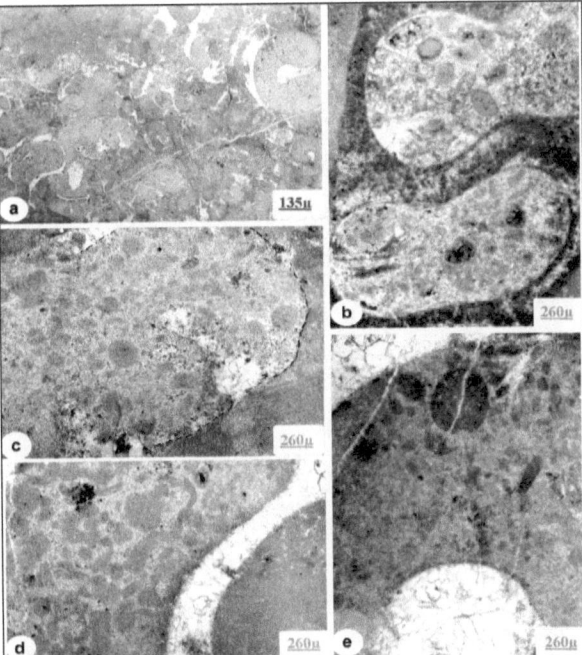

Fig.10: (a) Gastropod shell grainstone exhibits grain-supported fabric, moderate sorting and rudit size of whole and fragmented gastropods and bivalves, ordinary light (O.L).Figs. (b-e): Gastropod mouldic cavities filled with pelletal micrite, silty sized bioclasts, glauconite grains (Fig. b) and fine to coarse orthospars. P. L. (Figs. b & d & e) and C.N. (Fig. c).

2.4 Bioclastic peloidal packstone

The microfacies commonly interbeds with the terrigenous bioclastic wackestone-mudstone facies. Petrographically, the rock is Coarse-grained (80-520 µ) and poorly sorted. The microfacies consists essentially of peloids (> 50%) and bioclasts. The proportion of these allochems varies relatively to each others from one bed to another. The peloids comprise bahamite peloids (rotten bioclasts) that represent completely obliterated and micritized skeletal particles being mostly of micritic walled benthonic forams and/or algae and echinoids. Accordingly their shape and size are variable and inherited from the altered bioclasts.

The bioclasts are very abundant and diversified. They are represented mainly by echinoid plates and spines, bivalves and benthonic forams including Cuneolina, Dicyclina, Quniqueloculina (miliolids) and other planispiral and uniserial tests. Keeled and globular planktonic forams, planktonic filaments, calcitized radiolarian tests, bryozoan fragments and serpulids are also reported in this microfacies but being less common. Most of the recognized bioclasts are either calcitized and enveloped with thin micrite rind or partially to completely micritized and locally silicified (Fig. 11a).

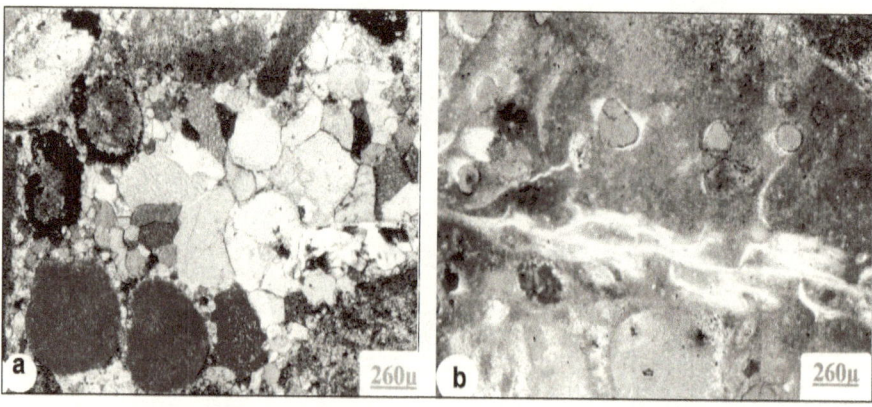

Fig. 11: (a) Patches of pseudo- and orthospars, bioclasts are either calcitized and enveloped with thin micrite rind or partially to completely micritized and locally silicified, C.N. (b) bioclasts mixed with honey yellow glauconite pellets and few ooids, P.L.

Many of the echinoid particles are enlarged by syntaxial calcite overgrowth. However, some molluscan shells still preserve their original wall structures. The bioclasts are occasionally mixed with honey yellow glauconite pellets and few ooids (Fig. 11b). The allochems are embedded in a dense algal micrite or argillaceous micrite with patches of pseudo-and ortho-spars (Fig. 11a).

2.5 Laminated peloidal foraminiferal grainstone

Allochem components and depositional fabric are nearly identical with the above-described bioclastic peloidal packstone microfacies. However, it is characterized by microscopic laminations consisting of very thin (25-110µ) discontinuous laminae of peloidal grainstone alternating with thicker (460µ) laminae of peloidal foraminiferal grainstone (Fig. 12).

The peloidal grainstone laminae, some of which are of single grain thick, are formed almost entirely of closely packed peloids with rare bioclasts and discrete patches of micrite matrix with scattered dolomite rhombs. The foraminiferal grainstone laminae are open packed and composed mainly of benthonic foraminiferal tests, being mostly of miliolids mixed with peloids and some fragments of bivalves, echinoid plates and algae. These are cemented by orthospars forming isopachous, granular mosaic and syntaxial texture.

Fig. 12: Laminated peloidal foraminiferal grainstone formed of alternated very thin discontinuous laminae of peloidal grainstone and thick-laminae of peloidal foraminiferal grainstone displaying microscopic lamination, O.L.

2.6 Bioclastic planktonic foraminiferal wackestone-mudstone

It is dominated with planktonic foraminiferal tests represented by trochospiral and globular forams (50% of total composition) associating with ostracod carapaces, delicate bivalves (pelagic) and planktonic filaments (Fig. 13a, b). Other biodetritus of echinoids, molluscans, algae, bryozoans and serpulids occur in subordinate amounts.

Texturally, the rock is fine grained, matrix supported and open packed wackestone and mudstone. It is commonly laminated and bioturbated . The laminae are of continuous and lenticular forms and comprise an alternation between mudstone and wackestone or between glauconitic rich (1300µ, thick) and glauconitic poor (650µ, thick) laminae. In the bioturbated beds, the burrow fill shows sharp boundaries and is of the same composition of the parent rock except being with abundant argillaceous micrite matrix.

The micrite supporting this rock type is generally argillaceous (Fig. 13a, b). In some beds, the micrite is partially to extremely neomorphozed into micro and pseudospars and also locally dolomitized to very fine and medium crystalline dolomite (range in size from 13-130µ).

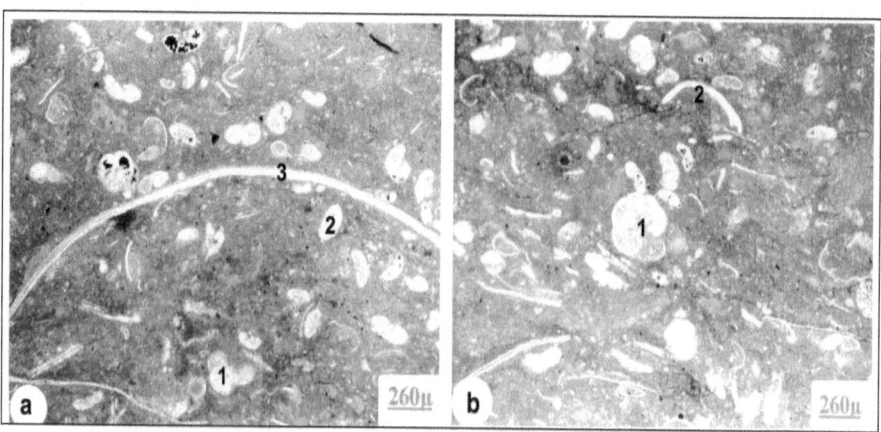

Fig. 13: (a,b): bioclastic planktonic foraminiferal wackestone consists of trochospiral (1) and globular planktonic forams associated with ostracod carapaces (2), delicate bivalves (3) and planktonic filaments (4), The micrite supporting this rock type is generally argillaceous, P.L.

2.7 Terrigenous bioclastic wackestone

The rock consists mainly of worn bioclasts being mostly of echinoid plates and spines with ostracod carapaces (Fig. 14a), benthonic and planktonic forams, delicate bivalves, algal and bryozoan fragments as well as scattered bahamite peloids and ooids. Angular to subrounded monocrystalline detrital quartz and glauconite pellets are also observed (Fig. 14b). The allochems of this microfacies are embedded in a dense micrite matrix being in some beds organic rich (Fig. 14a) and in the beds is glauconitic to slightly ferruginous. Partial dolomitization and neomorphism into micro and pseudospars are the main diagenetic features of the micrite matrix. Rock of this microfacies commonly displays bioturbation in both field and microfacies.

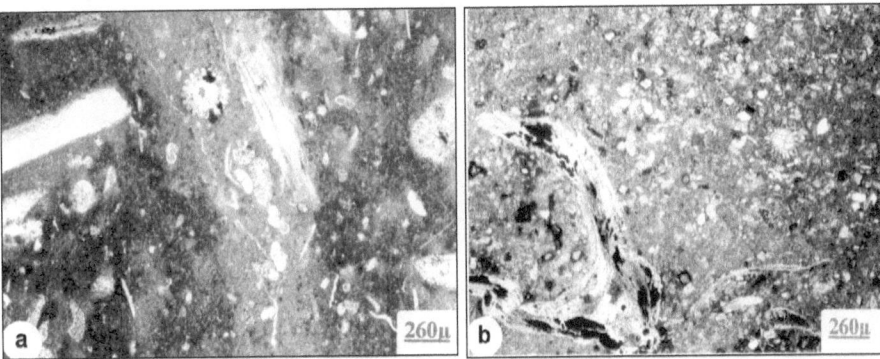

Fig. 14: (a) wackestone consists of bivalves, echinoid plates and spines, ostracods, planktonic foraminifers and delicate bivalves embedded in organic rich micrite matrix; P.L. (b) Terrigenous bioclastic wackestone consists of echinoid spines, delicate bivalves mixed with angular to subrounded monocrystalline detrital quartz. Notice Partial dolomitization and neomorphism of micrite matrix into micro- and pseudospars and partial replacement of some bioclasts by iron oxide, C.N.

2.8 Bioclastic whole echinoderm fossil wackestone

It is recorded only in the upper limestone-shale unit, where it forms two bodies of calcareous shale and glauconitic marl with thickness ranges from 1-1.5 m. The rock is essentially composed of large unfragmented echinoid plates mixed with scattered echinoid spines, ostracod carapaces and planktonic foraminifers

(Fig. 15a). These allochems are arranged in the echinoid plates and planktonic forams laminae. They are embedded in very fine calcisiltite micrite matrix.

2.9 Bioclastic ostracodal wackestone-mudstone

This microfacies is uncommon. It forms few beds (30cm to 100cm, thick). The rock is mainly composed of both articulated and disarticulated calcitized ostracod carapaces (~30% of the total composition) with subordinate planktonic and benthonic foraminifers (Fig. 15b). In different beds, the allochems are embedded in micrite matrix that varies in composition from bituminous to argillaceous (Fig. 15b). Microscopic bioturbation is observed in this microfacies. It is evident through different packing of ostracods between parent rock and burrow filling.

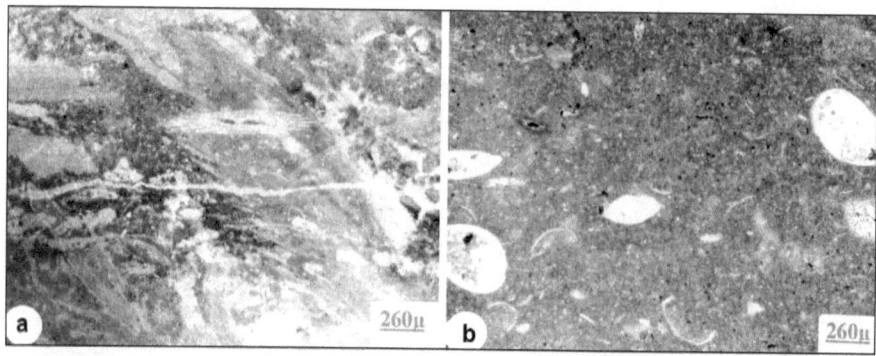

Fig. 15: (a) Large unfragmented echinoid plates (center) intermixed with scattered echinoid spines, ostracod carapaces and planktonic forams set in very fine calcisilitite micrite matrix, P.L. (b) Bioclastic ostracodal mudstone consists of calcitized articulated and disarticulated ostracod carapaces embedded in bituminous micrite matrix, C. N.

3. Facies Hierarchy and Depositional Model

3.1 Facies A: Bioturbated/massive skeletal wackestone-mudstone
Description

It forms the basal, finer- and deeper-water lithologies of several shallowing-upward subtidal cycles (Fig. 16 a & b). The facies is fossiliferous argillaceous limestone grading in few beds to calcareous fissile shales. The facies is rich in exotic grains of quartz and glauconite and displays various colors as light gray, chalky white, pale yellow to egg-yellow and yellowish green with general white burrow fills. The rock is fine-grained, moderately sorted and matrix-supported. Its macrofauna are abundant and diverse including gastropods, bivalves and echinoids. In representative thin sections, bioclasts of echinoids, benthonic forams, delicate bivalves, ostracods, calcispheres and serpulids are commonly observed with scattered bryozoan and algal fragments as well as occasionally planktonic forams.

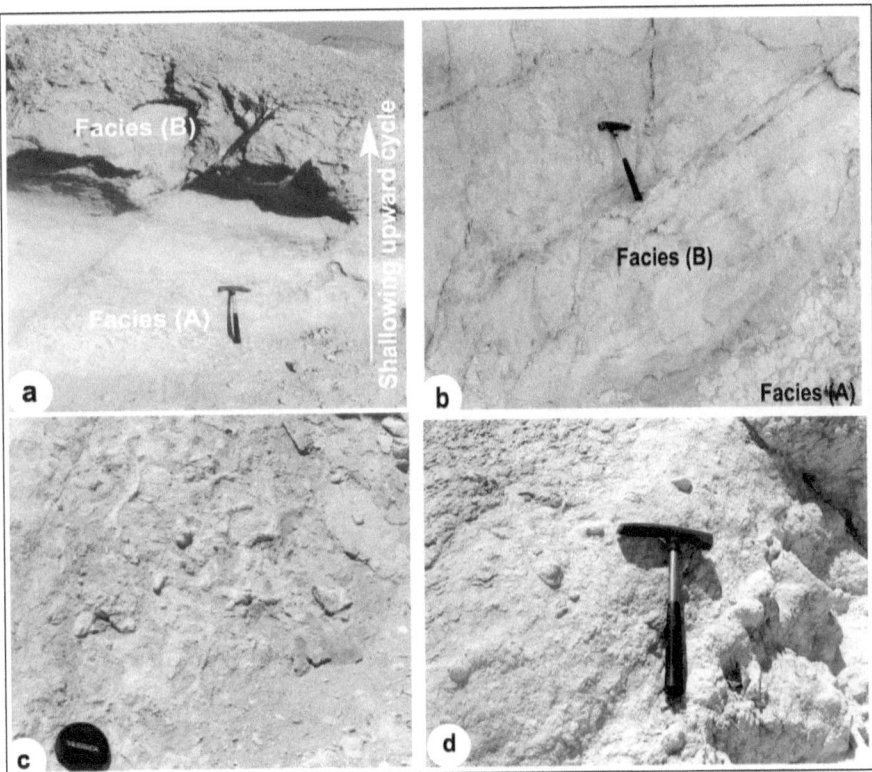

Fig. 16: (a & b): Shallowing upward subtidal cycles begins with basal finer and deeper water lithologies of facies A and ends with facies B. Notice the bed homogenization by intensive bioturbation. (c) Irregular branched burrows causing complete bed homogenization of facies A. (d) Nodular bedding or boudinage like structures characterizing subfacies A1 formed by intensive bioturbation and compaction.

The non-skeletal allochems are represented by scattered bahamite peloids, ooids with quartz, phosphate and glauconite grains. Sometimes, the facies has scouring sharp base on which locally reworked clasts are observed.

The beds of facies (A) are represented by the following microfacies association being in descending order of abundance: terrigenous bioclastic wackestone-mudstone; bioclastic ostracodal wackestone-mudstone and bioclastic whole echinoderm fossil wackestone.

In outcrops, the facies is almost made up of thinly to thickly flat-bedding (5-50cm, thick) with various internal depositional structures upon which the facies is subdivided into three subfacies types:

Subfacies (A1): Bioturbated/nodular-bedded skeletal wackestone-mudstone builds most beds of the slope-forming units (Fig. 3a). It is typified by intensive horizontal and irregular branched burrows causing either complete bed homogenization (Fig. 16 c & d) or with compaction give rise to distinct nodular or flaser bedding (sensu, Wilson and Jordan, 1983, Fig. 16d).

Subfacies (A2): Massive-bedded skeletal wackestone-mudstone forms very thick white chalky limestone beds being barren from any physical or biogenic structures, except occasionally thin levels of horizontal burrows near the base and the top of the beds (Fig. 3a).

Subfacies (A3): Laminated skeletal wackestone-mudstone is characterized by continuous to discontinuous planar fine lamination, locally with horizontal burrows.

Interpretation

The general fine grain size, mud-supported fabric, flat-bedded nature and lack of wave and current structures characterizing this facies indicate its deposition in a quiet water environment below effective fair-weather wave base (FWB). The intensive bioturbation and nodular-bedding in many beds or fine lamination in some others reflect a low rate of sedimentation in low energy conditions (Wilson and Jordan, 1983 and Wright, 1986). On the other hand, the internally massive beds may suggest a damping from highly concentrated sediment dispersions or rapid deposition from suspension (Prothero and Schwab, 1997). The distinct nodular and flaser bedding is very common structure in the subtidal shelf or ramp carbonates. Such structure is attributed to interplay of burrowing, selective cementation and differential mechanical compaction, particularly in inhomogeneous argillaceous carbonates (Clari and Martire, 1996). The diversity of fauna dominating with the stenohaline echinoids and some bryozoan fragments admixed with the euryhaline benthonic forams, mollusks and ostracods indicate open circulation of normal marine water (Heckel, 1972; Jones and Desrochers, 1992). The microfacies types constituting this lithofacies are comparable with the standard microfacies types 9 (Bioclastic wackestone-mudstone), type 10 (Packstone/ wackestone with coated bioclasts) of Wilson (1975) who assigned them to the standard facies belts 7, 2 (Shelf lagoon/ outer shelf). The presence of bioclastic packstone with coated and worn bioclasts reflects textural inversion, where the particles are formed in high-energy shoals and moved down local slopes to deposit in quiet water. The dolomitization in this facies is probably carried out in the early stage of diagenesis via seepage of Mg-rich brines as proposed by Adams and Rhodes (1960) and Muller and Tietz (1971).

Summing up facies (A) represents low energy, subtidal inner-or middle platform facies. Similar skeletal-pelletal wackestone facies is interpreted as foreshoal deposits accumulated in slightly protected subtidal environment in water depth 5-25m by Elrick and Read (1991).

3.2 Facies B: Well-bedded/bioturbated skeletal-peloidal packstone

Description

It currently caps the above-described facies (A), and terminates shallowing upward cycles (Fig. 16 a&b). The facies is typified by thin to thick (5-40cm, thick) even and wavy bedding (Fig. 17 a&b). It is made up of either chalky limestone or occasionally glauconitic limestone. Sharp soles and asymmetrical rippled fossiliferous tops characterize most of the beds representing facies (B). Almost those beds display internal planar or ripple cross-lamination and occasionally hummocky and trough cross-lamination. Horizontal or branched burrows are locally observed in some bases of those beds.

The facies is very fossiliferous with diversified fauna mostly of thick-micritic walled benthonic forams (e.g. miliolids, endothyrides, Dicyclina and Cuneolina and fragments of oysters, gastropods and echinoids that are almost worn with micrite rinds. Ostracods, algal particles, planktonic filaments, serpulids, bryozoans and calcispheres with scattered ooids and intraclasts are also encountered. Many of these allochems are rotten and converted to bahamite peloids.

Petrographically, the facies is essentially represented by bioclastic peloidal packstone and laminated peloidal foraminiferal grainstone. These microfacies types have mostly medium to well sorting, close packing and well-washed grain-supported fabric.

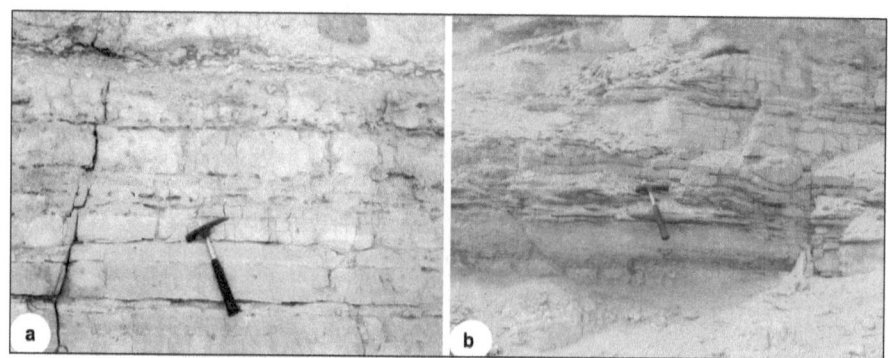

Fig. 17: Thin-to thick even and wavy bedded limestone of facies B.

Interpretation

Facies (B) with its characteristic grain-supported fabric, close packing, well-sorted nature and diversified faunal content indicate an accumulation in well-aerated open marine environment under high-energy conditions. The impact of effective wave and/ or current action on the deposition is evident from the pervasive horizontal lamination, rippled cross-lamination and asymmetrical ripples (Wilson, 1975 and Jones and Desrochers, 1992). The local preservation of wavy bedding and hummocky cross-stratification also reflects intermittent influence of storm wave or storm induced current and accumulation above storm wave base (Harms et al., 1975; Aigner, 1985; Leckie, 1988 and Simpson and Erickson, 1990). The benthonic forams dominating in this facies are assigned as typical restricted marine fauna by many authors (e.g. Murray, 1973 and Chaproniere, 1975) but, in contrary, other workers consider them as common assemblage in sand shoal setting with normal salinity (e.g. Brasier, 1975 a & b and Hallock, 1984). According to Carannante et al. (2000), these forams which belong to their foraminiferal assemblage β are typical of well-lit depositional environments characterized by an open water circulation, normal salinity and water depth at least 10m.

Based upon the above facies criteria, facies (B) is considered as prograding bioclastic foraminiferal bank or sand shoal facies. Similar bioclastic calcarenite facies is interpreted as skeletal grainstone shoal facies deposited in water depth 5-10m (Elrick and Read, 1991) and as subtidal mobile foraminiferal sand sheets (Carannante et al., 2000).

3.3 Facies C: Massive/laminated oyster shell grainstone (Oyster Hash)
Description

Oyster shell beds ranging from few cms up to 50cm thick occur at several levels, where it caps facies (A) and terminates the shallowing upward cycles. Oyster hash beds are argillaceous, partially dolomitized and mostly with erosive base and internal massive or faint laminated structures. It consists almost entirely of mono or dispecific oyster shells set in mudstone to wackestone matrix (Fig. 18). The majority of oyster valves show good preservation quality, almost disarticulated, slightly broken, little abraded and neither bioeroded nor encrusted. They have bimodal to fair sorting; mostly of complete valves with little comminuted debris. In side view, most shells are nearly aligned parallel to each other and oriented convex-up (Fig. 18) with closely packed fabric and stylolitic contacts that are filled with ferruginous clays.

In thin section, the facies is very coarse-grained, bioclastic-supported and locally shows a fining-upward grading. Scattered bahamite peloids, ostracod carapaces as well as fine-grained quartz and glauconite are recorded

Fig. 18: Close up view of oyster hash beds composed of oyster shells set in mudstone to wackestone matrix. Notice in Fig. a, the parallel alignment of shells, convex-upward orientation and fining upgrading of oyster shells.

Interpretation

Shell beds or shell concentrations are common features in both modern and ancient shelves. They are differentiated into several types on the basis of their biofabric, taphonomic signatures and formational processes (e.g. Kidwell et al., 1986 and Fürsich and Oschmann, 1993). In the study oyster shell beds, its distinct coarse-grain size, bioclastic-supported fabric and internal oriented nature of the almost disarticulated valves reflect an accumulation under relatively high-energy conditions. The good preservation with little fragmentation and abrasion of oyster valves and their setting in mudstone to wackestone matrix indicate minor reworking and winnowing or local transport and final concentration via short-lived events like storm waves or storm flows. Sharp to erosive soles and fining-up grading in those oyster beds (Fig. 18) strength the storm-influenced formation. These oyster hashes are almost comparable with the shell beds type 2 (storm wave concentration) and/ or type 3 (proximal tempestites of Fürsich and Oschmann (1993), which commonly occur in the subtidal environment below fair-weather and above storm-wave bases.

3.4 Facies D: Massive/bioturbated gastropod skeletal packstone-grainstone (Gastropod-shell beds)

Description

The facies occurs as thick to very thick grayish white fossiliferous limestone beds (Fig. 19) reaching in a cumulative thickness up to 10m (middle unit) but decreasing at El-Hassana dome to about 3m where the facies represents the substrate upon which the robust *"Durania arnaudi"* and the coralline sponge *Millestroma nicholsoni* accreted their mounds.

The gastropod beds are internally massive to highly bioturbated with egg yellow branched burrows (Fig. 20a). Its populated gastropod shells of *Trochactaeon salomonis* and *Nerinea requeniana* are good preserved (Fig. 20 b & c), slightly fragmented, almost unworn and being completely recrystallized. They are erratically oriented, moderately sorted and vary in population from dense bioclastic-supported to open packed matrix-supported. The matrix is of coarse packstone to grainstone fabric, moderately to well sorted and consists of worn bioclasts of diversified fauna comprising echinoids, algae, miliolids, cuneolinids, bivalves, ostracods and bryozoans admixed with bahamite peloids with few intraclasts and glauconite grains (Fig. 20d).

Interpretation

Coarse grain size, worn diversified bioclastic composition and grain-supported fabric of matrix with well bedded and bioturbated nature of the gastropod shell beds are diagnostic features for subtidal shoal or near-shoal facies accumulated under relatively high energy conditions (Flügel, 1982 and Wilson, 1975). The facies is comparable to the standard Microfacies Type 10 (coated and worn bioclasts in micrite) of Flügel (1982) who considered it to be formed in swales in proximity to shoals and subsequently moved down to be deposited in quiet water. The good preservation and little fragmentation of the dominant gastropods (Actaeonella and Nerinea shells) reflect that they inhabited and burrowed their grainy shoal or near shoal deposits during intermittent relatively quiet water conditions. However their concentration with erratic orientation could be attributed to sporadic storm wave action. Concentration of shells via storm wave is characterized by high preservation quality of shells and packstone to grainstone matrix (shell bed type 2 of Fürisch and Oschmann, 1993).

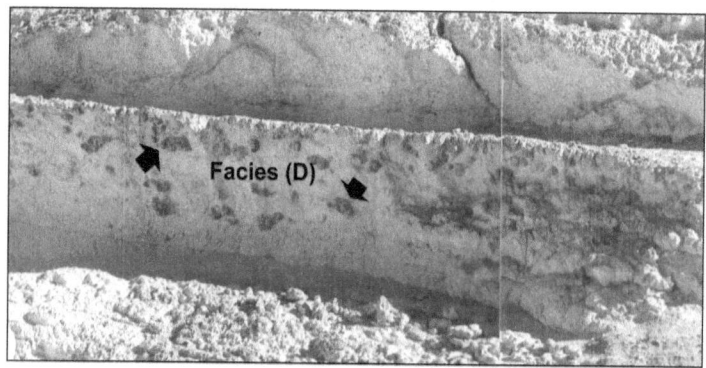

Fig. 19: Thick- to very thick grayish white limestone beds crowded with gastropod shells and coralline sponge heads (arrows), middle limestone unit.

3.5 Facies E: Massive rudist boundstone-bafflestone
Description

This type of rudist facies is best developed and good preserved at El-Hassana dome where it flanks its crest (Fig. 20e). The rudist facies changes laterally due NE into a thick unit (up to 10) of well bedded gastropod skeletal grainstone-packstone facies (Fig. 6) capped, in some places, by silicified coralline sponge boundstone.

The rudist facies assumes a small organic mound-like form (sensu Toomey, 1981) extending for about 133.6m in NW-SE direction (Fig. 20e). It is poor in obvious bedding feature and is built essentially by its own biota of the rudist "*Durania arnaudi*" that rarely occurs in the surrounding equivalent facies (Fig. 21a). Internally such rudist mass consists of autochtonous and parautochtonous limestone subfacies representing the successive stages of its growth (see Fig. 6), they comprise:
- Parautochtonous skeletal rudist floatstone (E1).
- Autochtonous coralline sponge-rudist boundstone (E2).
- Autochtonous rudist bounstone-bafflestone (E3).
- Parautochtonous skeletal peloidal grainstone (E4).

The parautochtonous bioclastic rudist-floatstone subfacies (E1) constitutes the base or substrate upon which the Durania shells and the sponge heads were attached and populated. It attains about 0.5m, consisting mostly of fragmented Durania shells (up to 10cm, in size) admixed with bioclasts of echinoid plates and spines, bivalves, gastropods, miliolids and peloids being set in white (occasionally iron-stained brown) micrite matrix. This floatstone substrate is overlain by autochthonous coralline sponge-rudist boundstone subfacies. The latter forms two irregular massive beds (up to 1m, thick, each) separated by a very thick bed (1.5-2m) of autochthonous rudist boundstone-bafflestone.

The autochthonous sponge-rudist boundstone (E2) is formed of closely spaced, very thick robust Durania shells and hemispherical heads of coralline sponge. Both members exist in a nearly equal proportion. They are good preserved and not fragmented, although their original skeletons are partially recrystallized and/ or silicified. The inter- and intra- skeletal voids are occluded with micrite, mosaic sparry calcite and occasionally by mosaic quartz. Laterally, (few hundreds of meters due SE), such boundstone facies forms local mound (up to 2m, thick) consisting almost entirely of the coralline sponge heads without any rudist shells (Fig. 21b). Most of the heads (up to 40cm, in diameter) are globular with numerous surface nodes as well as irregular and meandering shallow grooves. They are welded with white micrite matrix and sparry calcite.

The autochthonous rudist boundstone-bafflestone (E3) is the main rock type of the rudist mass. It forms a very thick massive bed (1.5-2m) that is overcrowded with robust shells of *Durania arnaudi* occasionally with few scattered small heads of coralline sponge, especially near the base. The rudist shells are very large up to 35cm in length and 8 to 15cm in diameter and are almost recrystallized. However, their external ornamentation in the form of radial costae and bands are still preserved. They are tightly packed with mutual contacts, and commonly inclined (Fig. 21a) or occasionally recumbent but some others instead are erect and in life position. In addition to the essential rudist component, the bafflestone subfacies includes echinoid fragments, benthonic foraminifers, little sponge fragments and peloids set in iron-stained micrite matrix.

The topmost bioclastic calcarenite subfacies (E4) attains 2 to 3m thick. It mantles the crest and northern flank of the rudist mound (Fig. 20e). The rock is packstone to grainstone, poorly sorted and ranges in size from medium to very coarse sand. At northern outcrops (200m north of the main mound), it displays large scale low angle tangential cross-bedding. Its allochems are essentially peloids and moderately rounded bioclasts including echinoid plates and spines, benthonic foraminifers, fragmented rudists and other mollusks, green algae and bryozoans. Many of the peloids and bioclasts possess secondary micrite rinds.

Interpretation

Rudist bivalves are a bizarre group of fossils that added immensely to the volume of calcareous matter in Cretaceous carbonate buildups (Wilson, 1975). They displaced corals as active builders for reefs and mounds during Cretaceous (Johnson and Kauffman, 1996). The restricted rudist mass and associated coralline sponge at El-Hassana dome are interpreted as reefal mounds by Hamza (1993) and as rudist biostrome by De Castro and Sirna (1996). Hamza (op. cit) attributed the restricted colonization of those sessile fossils to a presence of submarine highs resulted in response to a compressive tectonic phase deformed the Abu Roash area during the Turonian. However, upon the other domal structures that are adjacent to and identical with El-Hassana dome, the rudist populations did not develop, and its equivalent stratigraphic level is, instead, replaced by the *Actaeonella* bank facies.

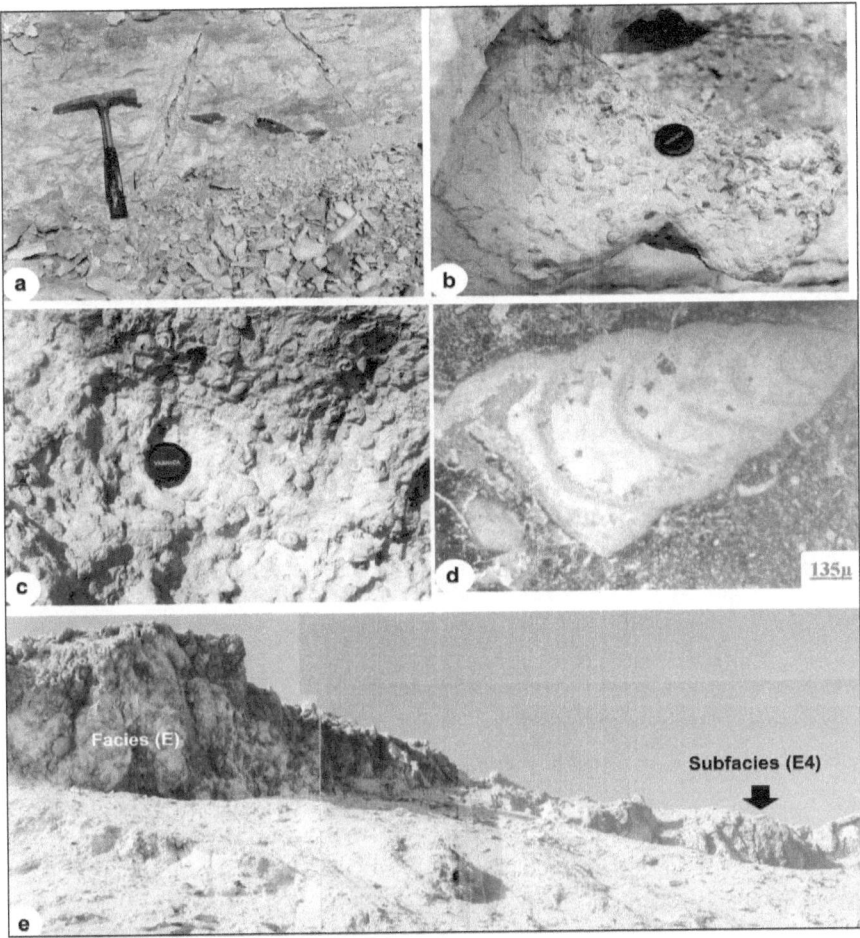

Fig.20: (a): Close up view of gastropod beds disturbed internally by egg-yellow branched burrows. (b & c) Close up view of gastropod beds rich with well preserved *Trochactaeon salomonis* and *Nerinea requeniania* shells. The shells are erratically oriented, completely recrystallized and slightly fragmented. Notice the beds in Fig. C are crowded only with *Trochactaeon salomonis*, middle limestone unit. (d) Photomicrograph shows well-preserved complete *Trochactaeon salomonis* shell sets in a bioclastic peloidal matrix, middle limestone unit, O. L. (e) Small organic rudist mound exhibiting poor bedding features.

It is known that, the rudist bivalves are not ranked as proper reef builders. Their individuality, olconal nature and the general lack of encrusters associated with them limited their ability to construct reef framework (Riding, 2002). The rigidity, relief, internal cavity development and other features of frame reef are not typical of rudist structures. However, the gregarious nestling by erect rudist forms (as the erect radiolitid Durania in the present case) as well as their fast growth, environmental tolerance and baffling ability, assist these forms, particularly in colonizing extensive parts of platform, from wave swept margins to protected interior parts. The colonized rudist masses formed of in situ closely packed radiolitids or hippuritids with elongate erect straight or gently curved valves are regarded as a typical example of Close Cluster Reef rocks (Grosheny and Philip, 1989). This reefal rock type is distinguished by in situ individual skeletons that are closely packed but not in contact, low topographic relief, storm disruption and scarcity of extra-skeletal cavities (Kauffman and Sohl, 1974). In some extreme cases (as in the study Durania mass), the very tight clustering resulted in mutual contacts between the adjacent rudist shells giving a subtle skeletal rigid framework. It thus places that rigid skeletal structure on the interface between Close Cluster- and Frame Reefs (Moro, 1997). Accordingly, the study El-Hassana rudist mass, particularly its main middle units of rudist/sponge boundstone-bafflestone subfacies (E2 & E3) represent a Close Cluster/Frame Reefal facies. Its builders of erect radiolitids (Durania) and hemispherical head of coralline sponges inhabited and colonized a soft bioclastic micritic mud that is preserved as irregular bands between the subfacies units. The presence of such fine grained bioclastic micrite and the good preservation quality of the reef builders reflect that the population occurred in shallow worm water and soft bottom protected from effective winnowing and destructive action of strong waves or current (Fürsich and Oschmann, 1993). According to Wilson (1975) the radiolitid forms, Durania and Sauvegesia were inhabited quiet water, while the other robust radiolitids were surf resistant and dominated in the outer shelf margin. The soft nature of the substrate supported stability of the upward growth of the erect Durania shells. However, their toppling and inclination may suggest intermittent storm action or is probably due to the load effect of the overlying sediments.

The above recorded succession of subfacies (E1-E4) constituting El-Hassana Durania mass is rather identical to the general growth sequence recognized by Wilson (1975) for many organic reefal mounds. The succession starts with rudist bioclastic floatstone subfacies (E1). Its matrix-supported and poorly sorted fabric as well as its predominate allochems of fragmented rudists and echinoids indicate an accumulation in a relatively quiet water; below fair weather wave base action. It is comparable with the basal bioclastic wackestone pile of most mounds sequences (Wilson, 1975). These basal micrite-supported piles are presumably heaped up by gentle currents and its accumulation is probably an important process in localizing mound growth. It represents the initial and start up stage of the mound growth. The rudist-sponge boundstone-bafflestone subfacies (E2 & E3) are equivalent to the bafflestone core and crestal boundstone of most mound sequence of Wilson (1975).

Fig. 21: (a) Close up view of the rudist mound overcrowded with closely packed and toppled Durania arnaudi shells. Notice the mutual contacts between shells, middle limestone unit. (b) Coralline sponge mound without any rudist shell forms the southeastern extension of the main rudist-sponge boundstone facies, middle limestone unit.

They constitute the thickest main part of the mass and represent the colonization and catch up stages of mound until reached effective wave base. Once mound top reached and remained for a considerable time at effective wave and current action, with stable sea level the upward growth and colonization ceased and fragmentation of mound top started. This is accompanied with accumulation of high energetic facies capping and flanking the mound. This stage is revealed by the topmost flat- and cross-bedded bioclastic peloidal grainstone subfacies (E4) possessing fabric and structure indicating high-energy conditions.

3.6 Facies F: Laminated/bioturbated planktonic foraminiferal wackestone-mudstone

Description

This facies forms few both calcareous mudstone and shale or chalky limestone beds, commonly intercalates with facies (A) (massive/bioturbated skeletal wackestone-mudstone) and forms together the lower and middle parts of the shoaling upward cycles (Fig. 22a &b).

Facies (E) occurs in thin to thick beds that are almost laminated or extensively bioturbated with nodular nature (Fig. 22c). The laminae are of continuous and lenticular forms representing an alternation between wackestone-packstone and mudstone laminae or glauconite rich and glauconite poor-laminae. The rock is fine grained, argillaceous, in parts glauconitic and almost matrix-supported. It is composed of the microfacies type bioclastic planktonic foraminiferal wackestone-mudstone. Planktonic forams of globular and trochospiral types are the essential fossil allochems and mixed with delicate bivalves, ostracods, echinoids, serpulids and bryozoan fragments (Fig. 22d).

Interpretation

The fine grain size, mud-supported fabric and laminated/bioturbated platy bedding nature reflect a deposition from suspension in very quiet water below normal wave base action. The predominance of the stenohaline planktonic forams with echinoids, bryozoans and delicate bivalves indicate well-oxygenated environment with normal marine salinity and good water circulation. The rock type of the facies is comparable with the SMF type 3 (pelagic lime mudstone of Flügel (1982). Its faunal content, depositional fabric and structures are of the common features characterizing open circulated platform (facies belt 2 or 7 of Wilson, 1975).

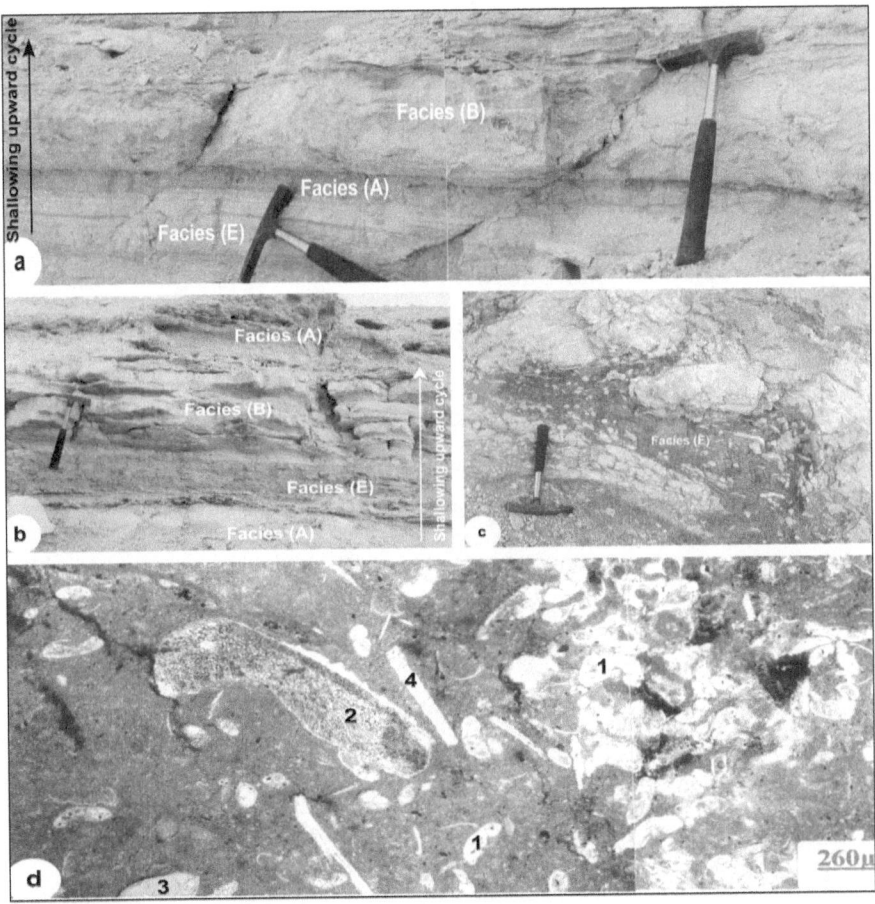

Fig. 22: (a &b) Shallowing upward cycles formed of facies E intercalated with facies A and terminated with facies B. Notice the laminated nature of facies E. (c) Bioturbated calcareous gray mudstone of facies E. (d) Photomicrograph of laminated/bioturbated planktonic foraminiferal wackestone-mudstone consists of planktonic foraminifers (1), echinoid plates (2), ostracods (3) and delicate bivalves (4) set in bituminous matrix, P.L.

3.7 Facies development and depositional model at El-Hassana dome

The facies analysis of the *Actaeonella*-bearing limestone-marl member in Abu Roash area revealed that its stratigraphic succession is formed essentially of two facies associations comprising open-marine subtidal assemblage and shoal or bank facies association (Badawy, 2003; Abu khadrah *et al.* 2014). The former association constitutes the lower and upper stratigraphic units of that member and includes the mud-supported facies (E) (Planktonic foraminiferal wackestone-mudstone) and facies (A) (skeletal wackestone-mudstone). Both of facies (E) and (A) are laminated to bioturbated, with abundant stenohaline fauna and are characterized, in this member, by an observable content of glauconitic, argillaceous- and in parts organic-rich micrite matrix. This type of coarsening-upward cycles is a common and distinct depositional product of storm-influenced middle to outer platforms (Read, 1985). The facies types (E and A) of each cycle represent the deposition from suspension in a quiet water condition that prevails in such depositional setting. While facies (B) and the molluscan shell beds are products of intermittent shoaling with reworking, winnowing and concentration via storm waves or currents.

The lower and upper stratigraphic units of that member are separated by a middle "*Actaeonella*" rich limestone unit (3 to10 m, thick). It is formed almost entirely of shoal/ bank facies assemblage represented by facies (D) (massive/bioturbated gastropod packstone-grainstone). Upon such gastropod shoals or bank facies, the robust-thick shelled *Durania arnaudi* with coralline sponge heads accreted local mounds in restricted areas (as in El-Hassana dome and western slope of El-Gaa Plateau). The Durania/sponge mounds comprise three main facies types representing the successive stages of its growth (Badawy, 2003). This facies includes: low-energy skeletal floatstone facies representing initial and start-up stage, rudist/ sponge boundstone facies of the catch-up and colonization stage, and high-energy bioclastic grainstone representing a cease of mound growth and destructive stage via effective wave and current action.

The abrupt development of middle to outer platform argillaceous carbonate facies with local rudist/sponge mounds of the *Actaeonella*-bearing limestone member above the inner platform shallower carbonate facies of the "limestone" member may suggest a retrogradation accompanied with a deepening of the depositional accommodation (Badawy, 2003; Abu khadrah *et al.* 2014). The sharp boundary between the two members could be considered as a transgressive surface. The enrichment of planktonic forams, pelagic delicate bivalves with abundant glauconitic and phosphatic grains near the base of the "*Actaeonella* bearing member may support such condition.

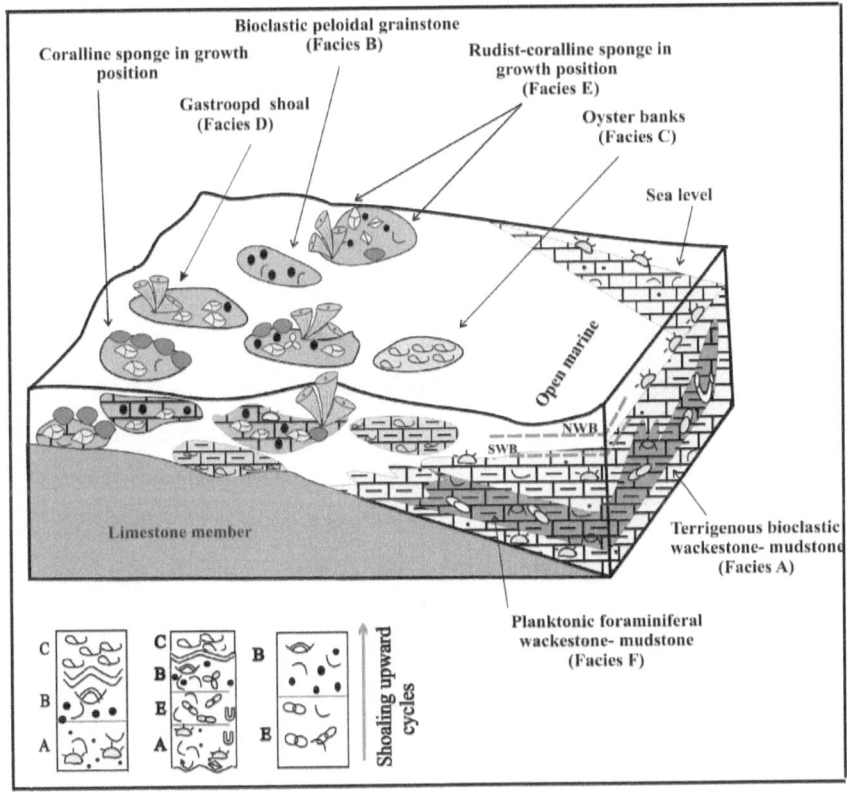

Fig. 23: Depositional model and shoaling upward cycles of *Actaeonella*-bearing limestone member at El-Hassana dome in Abu Roash area (modified after Badawy, 2003).

4. Concluding Remarks

Actaeonella-bearing limestone-marl member is best outcropped at El-Hassana dome and the western scarp of El-Gaa Plateau. It is recorded also in the summit of Gabal El-Ghigiga, Giran El-Ful, Gabal El-Hifaf and Tel El-Mabsuta. It exhibits a contrasting dark tone relative to the encompassing white cliff forming limestone-dominated members. It consists of shale, marl and limestone that yield in some places rudist shells with *Trochactaeon salomonis*, *Nerinea requieniana* and *Millestroma nicholsoni*. The *Actaeonella*-bearing member is subdivided into:

1. Lower limestone-shale unit is consisting of glauconitic argillaceous limestone grading to calcareous shale and intercalating with fossiliferous bioturbated limestone, flaggy bedded limestone and oyster hashes.

2. Middle (gastropodal) limestone unit consists of thick to very thick either massive to faintly laminated or bioturbated grayish white limestone beds containing disoriented whole gastropod shells of *Trochactaeon salomonis* and *Nerinea requieniana*. At "El Hassana dome", this gastropodal limestone unit is followed up by limestone overcrowded with large-sized *Durania arnaudi* and coralline sponge heads form local mounds. These fossils colonize in three to four irregular beds separated from each others by discontinuous layers of white bioclastic limestone containing skeletal fragments of rudists and echinoid. The rudist-coralline sponge beds are overlain and flanked by faintly cross-bedded and firmly cemented fossiliferous calcarenite beds. Very local and isolated patches of the large sized *Durania arnaudi* (Choffat) are also recorded for the first time in some of the upper beds.

3. Upper limestone-shale unit is consisting of yellowish gray to green calcareous shale inter-bedded with ledge-forming highly fossiliferous glauconitic argillaceous limestone. The calcareous shale is slope forming, fissile on the weathered surface, highly dissected with gypsum and salt veins and yields abundant delicate bivalves.

References

Abdel Khalek, M.L., EL Sharkawi, M.A., Darwish, M., Hagras, M., and Sehim, A. (1989): Structural history of Abu Roash district, Western Desert, Egypt. J. Africa Earth Sci., **9 (3/4)**: 435-443.

Abdel-Gawad, G. I. (1999): Biostratigraphy and facies of the Turonian in West Central Sinai, Egypt. Annals Geol. Surv. Egypt. **12**: 99- 114.

Abdel-Gawad, G. I. (2000): Coniacian gastropods from Sinai, Egypt. In: E.A.A. Youssef (Ed.) Geology of the Arab World, proc. 5th internat. Confer. Geol.Arab world, Cairo University: **3**, 1509-1526.

Abdel-Gawad, G. I. (2001): On some Upper Cretaceous coralline sponges from Egypt. Egypt. J. Paleontology, **1**: 299- 325.

Abu Khadrah, A. M., Abdel-Gawad, G.I., Helba,A.A. and Badawy, H. S. (2014): Facies Hierarchy of the Late Cretaceous Rudist-Coralline Sponge mounds and the Underlying Rudist- Actaeonella- Bearing units in Abu Roash Area, North Western Desert, Egypt. Tenth International Congress on Rudist Bivalves, Bellaterra, Barcelona.

Adams, J. E. and Rhodes, M. L. (1960): Dolomitization by seepage refluxion. Bull. Am. Assoc. Petrol. Geol., **44**: 1912-1920.

Aigner, T. H. (1985): Storm depositional systems. Lectures notes in Earth Science, Springer, Berlin, **3**: 174p.

Andrawis, S. (1990): Tables of foraminiferal biozones. In: Said, R. (ed.): The Geology of Egypt. 639-648.

Badawy, H. S. (2003): Stratigraphy and Facies Anlaysis of the Upper Cretaceous Abu Roash Formation in Abu Roash area, Giza, Egypt. Unpublished M.Sc. thesis, Faculty of Science, Cairo Univ., 191p.

Beadnell, H.J.L. (1902): The Cretaceous region of Abu Roash near the Pyramids of Giza, Egypt. Surv. Dept., Cairo, 48p.

Brasier, M. D. (1975a): Ecology of resent sediment- dwelling and phytal foraminifera from lagoons of Barbuda, West Indies. J. Foraminiferal Res., **5**: 42-62.

Brasier, M. D. (1975b): The ecology and distribution of resent foraminifera from reefs and shoals around Barbuda, West Indies. J. Foraminiferal Res., **5**: 193-210.

Carannante, G., Ruberti, D. and Sirna, G. (2000): Upper Cretaceous ramp limestone from the Sorrento Peninsula (southern Apennines, Italy): micro- and macrofossil association and their significance in the depositional sequences: Sed. Geol., **132**: 89- 123.

Chaproniere, G. C. H. (1975): Paleoecology of Oligo-Miocene larger Foraminiferida, Australia. Alcheinga, **1**: 37-58.

Clari, P. A. and Martire, L. (1996): Interplay of cementation, mechanical compaction and chemical compaction in nodular limestones of the Rosso Ammonitico Veronese (Middle-Upper Jurassic, northeastern Italy): J. Sed. Res., **66**, 447-458.

De Castro, P. and Sirna, G. (1996): The Durania Arnaudi biostrome of El-Hassana, Abu Roash area (Egypt). Geologica Romana, **32**: 69-91.

Elrick, M. and Read, J. F. (1991): Cyclic ramp to basin carbonate depositions, Lower Mississippian, Wyoming and Montana: a combined field computer modeling study: J. Sed. Petr., **61**: 1194-1224.

El-Shinnawi, M.A. and Sultan, I. Z. (1975): Lithostratigraphy of some subsurface Upper Cretaceous sections in the Gulf of Suez area, Egypt. Acta Geol. Acad. Sci. Hungaricae, **17**: 469-494.

Faris M.I. (1948): Contribution to the stratigraphy of Abu Roash and the history of the Upper Cretaceous in Egypt: Bull. Fac. Sci., Cairo Univ. **27**: 221-239.

Flügel, E. (1982): Microfacies analysis of limestones. Springer-Verlag, Berlin, 633p.

Fürsich, F. T. and Oschmann, W. (1993): Shell beds as tool in basin analysis: L the Jurassic of Kachchh, western India. J. Geol. Soc. London, **150**: 169-185.

Grosheny, D., Philip, J.(1989): Dynamique bio sedimentaire de bancs a rudistes dans un environment perideltaique: 1 aformation de La Cndiere d' Azur (Santonien, SE France). Bull. Soc. Geol. Fr. **5**: 1253-1264.

Hallock, P. (1984): Distribution of larger foraminiferal assemblages on two Pacific coral reefs. J. Foraminiferal Res., **14**: 250-261.

Hamza, F. H. (1993): Upper Cretaceous rudist-coral buildups associated with tectonic doming in the Abu Roash area, Egypt. N. Jab. Geol. palaont.Mh. H. **2**: 75-87.

Harms, J. C., Southard, J. B. and Walker, R. G. (1975): Structures and sequences in clastic rocks: SEPM Short Course No. 9, 161p.

Hataba, H. and Ammar, G. (1990): Comparative stratigraphic study on the Upper Cenomanian- lower Senonian sediments between the Gulf of Suez and Western Desert, Egypt. 10th Petrol. Expl. Prod. Conf., Cairo, Egypt, 16.

Heckel, PH. (1972): Possible inorganic origin for stromatactis in calcilutite mounds in Tully Limestone, Devonian of new York. J. Sed. Petr. 42, 7-18.

Johnson, C. C. and Kauffman, E. G. (1996): Maastrichtian extinction patterns of Caribbean Province rudists. In: N. MacLeod and G. Keller (eds.) Cretaceous-Tertiary mass extinction: biotic and environmental changes. Norton and Company, London: 231-272.

Jones, B and Desrochers, A. (1992): Shallow platform carbonates, in Walker, R. G. and James, N. P., eds., Facies models- response to sea level change: Geological Association of Canada: 277-301.

Jux, U. (1954): Zur Geologie des Kreidegebietes von Abu Roash bei Kairo: N. Jb. Geol. Palaont., **100(2)**: 159-207.

Kauffman, E. K. and Sohl, N. F. (1974): Structure and evolution of Antillean Cretaceous rudist frameworks.. Naturforschende Gesellschaft zu Basel, Verhandlungen, **84**: 399-467.

Kidwell, S. M., Fursich, F. T. and Aigner, T. (1986): Conceptual framework for the analysis and classification of fossils concentrations. Palaios. 1: 228-238.

Leckie, D. A. (1988): Wane formed coarse-grained ripples and their relationship to hummocky cross-stratification: J. Sed., Petr., **58**: 607-622.

Moro, A. (1997): Stratigraphy and paleoenvironments of rudist biostromes in the Upper Cretaceous (Turonian-Upper Santonian) limestones of southern Istria, Croatia: paleogeography, paleoclimatology, paleoecology, **131**: 113-141.

Muller, G. and Tietz, G. (1971): Dolomite replacing cement A in biocalcarenites from Fuerteventura, Canary Islands, Spain. In: O. P. Bricker (ed.), Carbonate cements. Johns Hopkins University Press, Baltimore: 327-329.

Murray, J. W. (1973): Distribution and ecology of living benthic foraminiferids, Crane, Russak and Co, New York.

Omara, S. (1953): The structural features of the Giza Pyramids Area. Ph. D. Thesis. Fac. Sci., Cairo Univ., 85p.

Osman, A. (1954a): Facies analysis of the Mesozoic surface and subsurface Formations of Abu Roash based on the percentage of characteristic microfaunal families and genera: Bull. Inst. Egypt, **36**: 177-180.

Osman, A. (1954b): Microstratigraphy of the Upper Cretaceous surface formations of Abu Roash: Bull. Inst. Egypt, **36**: 181-191.

Prothero, D. R. and Schwab, F. (1997): An introduction to sedimentary rocks and stratigraphy. W. H. freeman and Company, New York. 575p.

Read, J. F. (1985): Carbonate platform facies models. Amer. Assoc. Petrol. Geol., **66**: 860-878.

Riding, R. (2002): Structure and composition of organic reefs and carbonate mud mounds: concepts and categories. Earth Science Reviews **58**: 163-231.

Said, R. (1962): Geology of Egypt. New York, Elsevier, 377p.

Sehim, A. A. (1986): Surface and subsurface geologic investigation of Abu Roash area, Western Desert, Egypt. M. Sc. Thesis, Cairo University.

Simpson, E. L. and Erikson, K. A., (1990): Early Cambanian progradation and transgressive sedimentation pattern in Verginia: an example of the early history of a passive margin. J. Sed. Petr., **60**: 84-100.

Tewfik, N. and Ebeid, Z. (1975): On the stratigraphy of Upper Cretaceous in the Gulf of Suez Region, Egypt. Revista Espanola De Micropaleontologia, **7 (3)**.

Wasfi, S., Hassouba, M., Mokktar, A., Badawi, A., Shafy, A.A., Azazi, G. and Sakr, S. (1986): Abu Roash C sandstone environmental model for hydrocarbon exploration, EGPC 8th Expl. Conf., 31p.

Wilson, J. L. (1975): Carbonate facies in geological history. Springer-Verlag, Berlin, 471p.

Wilson, J. L. and Jordan, C. (1983): Middle shelf. In: P. A. Scholle, D. G. Bebout, and c. H. Moore (eds.), Carbonate depositional environments; A. A. P. G. Mem., **33**, 297-343.

Wright, V. P. (1986): Facies sequences on a carbonate ramp: the Carboniferous limestone of Wales: Sedimentology, **33**, 221-241.

Acknowledgments

The present work was financially supported by the Faculty of Science, Beni-Suef University.

Great thanks are due to Prof. Dr. Ahmed M. Abu Khadrah (Prof. of sedimentology and sedimentation in Geology Department, Faculty of Science, Cairo University), Prof. Dr. Gouda I. Abdel-Gawad (Dean of Faculty of Science and Prof. of paleontology and stratigraphy in Geology Department, Faculty of Science, Beni-Suef University) and Dr. Adly A. Helba (Lecturer of sedimentology and sedimentation in Geology Department, Faculty of Science, Cairo University) for proposing the present point of research, continuous help, reading critically the manuscript and fruitful discussion during the progress of this work.

Special thanks are due to my husband Prof. Dr. Mohamed Shahien (Prof. of applied mineralogy in Geology Department, Faculty of Science, Beni-Suef University) for his help during the field work and continuous support.

I want morebooks!

Buy your books fast and straightforward online - at one of the world's fastest growing online book stores! Environmentally sound due to Print-on-Demand technologies.

Buy your books online at
www.get-morebooks.com

Kaufen Sie Ihre Bücher schnell und unkompliziert online – auf einer der am schnellsten wachsenden Buchhandelsplattformen weltweit!
Dank Print-On-Demand umwelt- und ressourcenschonend produziert.

Bücher schneller online kaufen
www.morebooks.de

OmniScriptum Marketing DEU GmbH
Heinrich-Böcking-Str. 6-8
D - 66121 Saarbrücken
Telefax: +49 681 93 81 567-9

info@omniscriptum.com
www.omniscriptum.com

www.ingramcontent.com/pod-product-compliance
Lightning Source LLC
Chambersburg PA
CBHW031552210526
45464CB00003B/1271